INTELLIGENT UNIVERSE
What is the future of life in the universe?

WLADIMIR MOREIRA DIAS

UNIVERSE
INTELLIGENT

"when we speak of infinity, there are no eternal facts, as there are no absolute truths."

Copyright © 2014 by Wladimir Dias

6

INTELLIGENT UNIVERSE
What is the future of life in the universe?

This book is an english version of Universo Inteligente, the portuguese original edition published in Brazil by Ag Books. Copyright © 2014 by Wladimir Dias. This edition was prepared by author in consultation with G4 Editorial – Santos - SP.
Editorial production – Kiev Publisher
Catalogação Internacional de Publicação (CIP) (Câmara Brasileira do Livro, SP, Brasil)
Índices de catálogo sistemático: 000 1: 000 000 - 2014
IMPRESSO NO BRASIL - PRINTED IN BRASIL

"This book is dedicated to revolutionaries and passionate about issues involving the life and universe."

Thank you the collaboration and
confidence of everyone who helped
me make this project a reality.

what is the ultimate fate of our universe?

PREFACE

This book describes a cosmic journey through what ways the universe could reach the point of self-awareness, featuring unpublished ideas, incredible possibilities in the area of computing and nanotechnology, planetary engineering and different forms of life on other celestial bodies. Faced with the search for alien civilizations, quantum computers that can overcome our intelligence and nano machines capable of establishing artificial life through the Cosmos. In addition to all this content several boxes with trivia and callouts is comprised of pages, offering a more dynamic reading and always seeking the details (subtext) of life.

If you read it, prepare to receive ideas that can do you surely rethink their belief system about the universe and life. In a materialistic world and reductionist, stand before arguments coherent and courageous that lead us to consider the possibility of the Cosmos and consciousness like the same thing, it's certainly a privilege that will expand your mind greatly. As Einstein said: "a mind that opens to a new idea will never return to its original size" Good read!

Além de todo esse conteúdo diversos boxes com curiosidades e textos explicativos entremeiam as páginas, oferecendo uma leitura mais dinâmica e didática buscando sempre as entrelinhas da vida.

Se for ler, prepare-se para receber idéias que poderão fazê-lo certamente repensar seu sistema de crenças sobre o universo e a vida.

Num mundo materialista e reducionista, estar diante de argumentos coerentes e corajosos que nos levam a considerar a possibilidade do Cosmos e da Consciência serem uma só coisa, certamente é um privilégio intelectual que irá expandir sua mente sobremaneira.

Como Einstein disse: "uma mente que se abre a uma nova idéia jamais voltará ao seu tamanho original". Boa leitura.

PRESENTATION

The complexity of creation, in all its aspects, has complicated theories and causing warm discussions on these topics that fascinates us all so much. Today with the extraordinary advancement of nanotechnology and the mapping of the human genome, it is possible to get some answers and get closer to the final frontier, the basic bricks of that all the universe would be formed. This book will give you an idea fairly wide on all aspects of the evolution of the early universe, involving the first biological cell, the extinction of the dinosaurs, the human nature, the genome, the artificial intelligence, the conquest of space, the quantum computer and certainly will help to clarify many of the ideas of the reader, giving him conditions to search for yourself, an enhancement of the themes discussed here.

In this our great search for answers, there are a group of researchers considered little traditional, always trying to find some kind of State of Ecstasy based on your scientific evaluations, where it would be possible to access the call " exponential life". When I am visiting their laboratories I like listening to his theories and concepts about life and the universe. Generally speaking, I think we should always be open the new ideologies existentialists, that somehow helps us to assemble this puzzle of our true origins.

I was very curious on the subject and was soon asking, what exactly does this term "exponential life" because I had no idea what it was about. One of the researchers then after serving yourself with a little liqueur, kindly was explaining myself, telling me it was a kind of expression of life, whose structure would be composed of various dimensions, that multiply exponentially, always based on different frequencies, forming a kind of endless conexion between these universes, parallel, superimposed and intertwined, but that could not be perceived by normal senses.

This would happen because their shades are a little different, in fact, some of these theories are well explained by the fundamentals of quantum physics, when it allows to understand the behavior of micro particles within the microcosm of existing energy within the internal structure of the atom, I don't know if you know something about this subject, but it's pretty interesting. The energy inside this universe, there is no any consideration about space and time and these particles change their behavior when somehow feel they are being watched. I know a bit about this subject told him and so after this short breaks continued...

We believe that these universes exist, however, cannot be observed but can only be felt and in a remote possibility perhaps subtended, however rarely we get to make some kind of contact. For this, we use some specific techniques of transcendences, based on the vision that we have on the subject as a whole. The matter of the universe is considered by us a simple expression of energy, this definition also applies to all other parallel universes and specifically in our case, this condition starts on the basis of the internal structure of protons, which is one of the components of the atom.

For us yet, this energy expression oscillates at specific frequencies, being unique to each universe, setting up so this way, various characteristics of matter and in no event, thinking in hypothesis if some of these fundamental frequencies could be changed, would be unimaginable everything with amazing transformations in the form of matter as we know it, transposing all the fundamental concepts about space and time, opening up the opportunity for possible close encounters with other dimensions and for this, we live searching in perfecting our deep insight exercises, seeking some transition sectors between the abstract of life and mind.

In summary, we believe that these quantum fluctuations amplified determined the history of our universe, the solar system, the planets and still on a genetic level, determining the existence of complex adaptive systems and still the man himself.

Please, analyze with me, the logic, because if everything in our lives has a certain vibration frequency, including our thoughts and emotions, fructifying in a reality formed in principle by a cluster of atoms, which vibrate in different ways in various States of energy. Imagine with me, if the man in an exercise of intelligence, which is very peculiar, can in the future, through the possession of details of these dimensional features of matter, ignore this information, not using them to construct their reality the way he want.

I'm sure they will do it, said to him emphatically and then ratifying. I think this is really possible, considering that our perception of reality, depends on just the way our brain processes the information it receives, through our senses, all this would be like a tune of radio, that the change of frequency, we are also changing the music being played.

When we enter this cosmic rhythm of pulse, is relevant also understand the need that life has of pulsar in its various prisms or sides without getting stuck on just one of them and everything that exists through it, has a bit of cosmic truth, which is not absolute, because to our reality, the absolute does not exist only relative. The absolute is infinite and grandiose enough to be contained in only a concept, an image, since it involves infinite points where it flows, but to the relative of our temporal life, these movements are resonance pulsatórios are present from the moment of our birth, for it is he that differentiates an organism on the other and the unify in himself.. When this tuning happen, if instantly creates a grandeur and a stream of life that runs through the soul. She creates the melody for the contact and organizes the dynamics of the individual, acting through an integration that holds steady, until the moment of separation that is the culmination of this process, defined by the miracle of birth.

This pulse is directed to the same end and this involved a same fate, in a process of harmony between all, in a dance of encounters, which occurs from all lives, which is absolute in its essence. The riddle of life, for us works like a mountain range of hopes, where through the joy of exist even for a moment, can lead us to realize our desires of eternal life. We can never live from a manner correct and smooth, until we know what is the reason of our lives, why we exist and what exactly we came We came to do here as people.

We are extremely adaptable to vicissitudes of environment and even the most rudimentary, always we are interacting and somehow modifying it, in our continuous search of energy for the survival of the species. Our mechanisms of adaptation work always to prevent collapses and we guarantee the organization of systems for life. For example, if we leave our room at random, in a little time we will be living in a real mess, with dirt and objects by all sides.

In this situation, someone will have to perform some work, to provide the energy required, so that the room back to its previous condition of organization, otherwise, he will never spontaneously return to its initial state of organization. This is a reversible process, different from the aging of a person, which is an irreversible process, where there is no possibility of returning to the initial state even with external work. The most of our time is spent unconsciously seeking what pleasure us, acting through our thoughts, emotions and especially of our cognitive ability. This lifespan can be regarded as a scalar field defined in the entire universe and with constant variation in a single direction, always increasing at different rates according to the region of space in which it is and its speed in relation to the Supreme referential. We believe that life's goal is to search for the harmonization of all energies and in this process, our soul will be the interface, having an essential role, because it would act as a deposit of these energies, that would build up during the course of our lives.

We believe that there are various types of energy, and this energy wouldn't be specifically generated by the world of the atom, but would be of a different kind, with much more the abstract form and features that could only be developed from the energy used for transformation, a direct consequence of our actions. If this action is altruistic, would adding as an aura, but if it's a selfish action, The total stored energy will be smaller than before. We also believe that all of us when we were born, we received a default value of this energy, always set by the average energy of all living creatures and the from this moment, this oscillation would depend on the actions of this individual in the course of his life. This means ... For us, this energy level will influence directly on our incorporation into other levels of life, in addition to defining our spiritual level in this new location. For life, this level would act directly in your final result of evolution.

For them, she has an undeniable characteristic of transformation and thus will penetrating in all the interconnections of the existence. We are all a link that participates in this development and we are continuously contributing in this process of accumulation and Depuration of energy. The great purpose of life is the conquest of omniscience and omnipotence, thus changing the essence of us all, which today is only relative, and temporal.

But, to our reality, what really matters is the intensity of the moment. Our life is worth not only for its duration, because one life brief, also has its value in herself, getting set in the intensity in which we live every moment that is given to us and we can somehow give our existence a new color, a new flavor and a whole new meaning. It is valid mainly by the intensity we give, by several goals that animate and depending how we act, we are forming a better future for all.

Our mind is a combination of feeling, perception, idea and consciousness. Hence the importance of seeking at this time, the wisdom and understanding to practice acts that provide an evolutionary path for the life and non-regressive. I like a lot of these topics and some time ago I read an article suggesting to the reader, that in order to understand our universe, we could imagine us too far , around a trillion light years from Earth and continued ... In this event, this article considers that our universe seen from this distance, it wouldn't be more than a point with light twenty billion years and that could perhaps to coexist jointly with other universes, similar or different. In fact, it would be as if it were a single cell in a larger body represented by the infinite.

I found very interesting this definition, even quite simple, considering the complexity of the details involving our lives. Really would be wonderful if in the future, all of this were true, because it would open a fantastic range of research and possibilities, at the moment unimaginable to our reality.

Maybe include inside a chance too much remote , to change until all our references on the space-time. The great truth is that unfortunately our science is still far from solving the enigmas that surround us, even we still need to break up the speed of light barrier, but the difficulties of principles are part of the process of human evolution. Considered a result of our long Including in a chance well, remote change until our references on the space-time. The great truth is that unfortunately our science is still far from solving the enigmas that surround us, even after we broke up even the speed of light barrier, but the difficulties of the principles are part of the process of human evolution. That can be considered a result of our long evolution genetics, who added the personal trajectory on a physical medium, has defined who we are today

When I look at the stars, I understand, how much we human beings, we know nothing, we live still shrouded with many enigmas about our true reality, if we are really just a link in time, or if we are more. We haven't been able to answer basic questions, such as where we came from and where we're going, or how did the first cell and if we are alone in the universe.

These are intriguing issues of our lives and sincerely I believe that a great deal of human suffering throughout history happened mainly due to this little understanding we have of our true origins.

I find this fascinating subject, especially with regard to the creation of the first cell. I always try as far as possible, form concepts about everything that relates to life, but always focusing on more, their influence on the emotions, which in my opinion, is the true light of the matter, what really matters in life. I think all these questions about creation are very intriguing and it requires a high degree of subjectivity in its conclusions because it depends on a lot of the vision that every person has of itself and the elements that make up our universe.

Looking for my friend and researcher after a brief smile, I asked her if he had a few minutes to listen to some of my theories about the creation, but obviously, without any commitment to the truth of the facts, are just very particular ideas and concepts. Right now, he sat on the edge of the bed and told me: of course! I'm all ears. Well, I think in this enigmatic context of evolution, the emergence of the first cell really is one of the most intriguing and a few years ago, I felt the need to form some ideas about these concepts, because these are issues which are intrinsically linked to our emotional, because it refers to our origins.

Objectively, to me she is the result of combining the nucleic acid which is an extraterrestrial element that formed in deep space and were brought here by the meteors, with the existing elements on early Earth, as nitrogen, oxygen, resulting in a combination very simple and that were always in casual situations within the environment where always happened new contacts and recorded sequentially via a chemical behavior of your set, always on the basis of variants which occurred, as possible electrical discharges, the thermal variations and many other factors and that was offering all conditions so that through their records, seeking their best behavior for evolution in relation to the environment, but always keeping his best result for its evolutionary line, acting like a true software of life, using the several gathering data during billions of years before it was possible its creation, the result of a simple meeting of countless chemical records.

Think with me, if since the most remote principles of the existence, the ultimate goal in life has always been harmony between the elements, would be something like a continuous trend towards the unification of all apparent opposites that are part of our lives and that actually act only as adjuncts, because our essence is unique and immutable. It's funny when it examines some aspects of life, which although seems like a transparent simplicity in its principles, is very complex, because it has many conflicting trends. At this moment i was interrupted by the friend researcher , listening that had entered in a theme who considered his specialty because it was somewhat philosophical and of course I listened carefully to your opinion or insight into this subject under discussion. He then continued ... I agree in part with you, really I think that life has many plans and trends, I think also a mistake until very common, when we take a wish too far, at the expense of everyone else, because life is like you said, is basically a set of trends that need to be organized more coherently as possible, to facilitate this our continued search for our balance emotional.

After all, we always have in our mind, a multitude of desires, conflicting and often intertwined. In it there are different primitive impulses, rational, egoísticos, altruísticos and all need to be very well managed, always seeking as ultimate goal, the intelligent harmony between the elements, which are expressed primarily through our emotions and also considering the of other beings. This is the operating principle of any concept of harmony. Now the life is simple at its core, we just need to explore its sensibility and understand their details, but to get the success it becomes necessary, the use of the intellect, seeking the simplest principles, which really isn't very easy. I agree refutei, but anyway, no matter what, any context that exists in the universe, I think it's great that we can participate of this process of life, at least for a moment, considering that all us we are eternal in our desires.

For me, life is only a perception and all the difference this simply in how we choose our lenses through which we see the world. Within the context of evolution of life, we know that our universe is expanding rapidly in rotational movements, translational and helical a demonstration of fantastic balance, which if viewed mathematically would look a real mess, but with incredible accuracy in the balancing of the hundreds of forces involved in every microsecond of transition. However in a distant time. According to some astrophysicists, this inflationary behavior can lead us to two destinations anything pleasant. The first would be resulting from continuous expansion, where all matter and the heat would scatter throughout eternity. The second target would be a possible contraction, i.e., after expanding for billions of years the fabric of our universe would be contracted by gravitational attraction until it gets close and collapse. In other words, our order can be the eternal cold or heat up to a cosmic crush.

One way or another, will be a sad fate. Behold within this complexity would dare imagine a third and surprising alternative: the universe could generate another universe before his tragic end and, for this, he would need to improve and evolve their own conscience. This would be spectacular, and not impossible, because in principle this our gigantic universe is favorable to life and therefore at some point life forms would develop at different points of their vastness. From there, the evolutionary process would lead to the establishment of species and intelligences increasingly cleared that could reach a level of consciousness that would question its own existence. Gradually, these organisms could gather knowledge, develop science and develop more technologies and more sophisticated until they can transfer the intelligence to machines, as we do today with our computers.

Ideally, from these demonstrations of forms of life, be they natural or artificial intelligence our universe could create conditions to spread, to enter into communion and expand until the point at which it would take some kind of awareness about himself and about its complex operating mechanism, allowing the creation of a new universe. This idea into a superficial analysis would seem very crazy and highly risky, but found defenders among big names in the current science.

It is clear that otherwise, the future of life is not in this universe or in this dimension, namely, the man will end up inside of a fantastic and unknown quantum universe formed by sub-particle with energy, that certainly will introduce us at the future many of the laws that will enable the continuity of life regardless of matter.

I believe these responses will be far beyond the so-called God particle by boson, where scientists would meet laws and logical principles that let us further from the material life and much closer to the abstract life, leaving us more incredibly independent of action defined by the atom in our universe. We would simply can act as the neutrinos already known by science, but of course in a way that allow to exist the life and consciousness.

THE BEGINNING
UNDERSTANDING THE EVOLUTION
THE UNIQUENESS OF THE SCORPIONS
THE MOST EXOTIC OF CREATURES
THE BEAUTY OF THE CRYSTALS
PARABLE OF THE TIME
THE HISTORY OF FIRE
ANCIENT CIVILIZATIONS
LEONARDO DA VINCI AND OTHERS
THE APPEARANCE OF THE FLOWERS
GENETIC IMPROVEMENT CYCLE
THE EXTINCTION OF THE DINOSAURS
ARTIFICIAL INTELLIGENCE
ALIENS
THE LAST FRONTIER
HUMAN NATURE
A LIGHT AT THE END OF THE TUNNEL
END

UNIVERSE INTELLIGENT

O The universe is a constant blowing of wonders and life allows us to feel them and admire them, but if I can understand them without myths is the foundation that underpins and determines the meeting of man with the world. Today almost all scientists agree that the universe originated between fifteen and twenty billion years ago, resulting in a gigantic explosion, popularly known as the Big Bang theory, being considered as one of the most beautiful intellectual achievements of recent decades. For their development have a lot of contribution of the science involving the macrocosm, represented by the Cosmology and Astrophysics and science of microcosm, represented by subatomic Physics. . After the Big Bang, matter and energy were distributed in astonishingly uniform and all regions of the universe were born of this explosion, in the same time and exactly with the same force, only a tiny portion of diversity occurred in a paradox, enabling the formation of galaxies and of the systems. Constituted basically by atoms of hydrogen, helium, oxygen and carbon, representing ninety-nine percent of everything. Our via Lactea began to form about ten billion years, when the first embryos of stars appeared, formed by condensation of hydrogen, which in the midst of thermonuclear reactions constants, was turning into other elements, first the helium, carbon, which were then if combining and provoking new reactions, allowing the birth of the first stars and also the release of an amazing amount of energy into space in the form of light and other electromagnetic radiation, that have spread across the Galaxy, forming among others the forming among others our Sun, which is millions of miles, but even so, we still get a tiny portion of their energy, which doesn't get lost in the void of space and comes to bring us life.

No wonder the primitive man has worshipped above all things, in a ritual that involved respect and fear in the face of energy so powerful and unknown, in their belief, could only from a God. Just recently, we have begun to know something about your facet of fire and we now know, that your energy is the result of massive Atomic reactions, in demonstrating that neither he or any other star are eternal, but has its life cycle determined by the amount of fuel available to feed their ongoing nuclear reactions

Today we seek in a mixture of fascination and practical spirit to better know the space that surrounds us and that becomes increasingly important for the progress of our planet. Its composition is a little different, being silicon, oxygen, aluminum and iron, and can be considered a cosmic anomaly, but an anomaly with life in all its corners, where always manifests some kind of micro-activity, which may be in the most torrid of effluents or in the most frigid of peaks, life exists everywhere, sliding, crawling, walking, digging or swimming.

Even microbes are far from being stupid, because they are able to learn from the experience. There are bodies that see under ultraviolet light or blind who perceive the environment involved in an electric field, some living beings only an hour, other generous thousand years, doesn't matter, the fact is that all live in full harmony with the environment, representing the same life.

There are superficial differences that understandably strike us as important, but deep inside the heart of life, all of us, we have a bit of sequoias and nematodes, viruses and eagles, mud and humans, nearly identical, enjoying this fantastic system natural complex so integrated within us, only possible due to the action of the nucleic acid, the master molecule of life and most probably formed from the combination of organic compounds brought by meteors and very simple hydrogen-rich materials found on Earth and with the crucial and from molecular building blocks, determinant in the common characteristics of all living things, after all the evolution of life on the planet, based on this same organic molecule. The enigmatic context of evolution, the emergence of the first cell appears as one of the most intriguing of the puzzles, but it is believed that when the molecule of nucleic acid if formed, began to combine with the existing elements on the planet, such as nitrogen, oxygen, carbon, resulting in very simple and combinations that were always so casual within the environment, but the each new contact, was always recorded via chemical behavior sequence of your set, on the basis of variants which occurred, as possible electrical discharges, the constant thermal variations, the continuous exposure to various chemical situations and many other factors, giving conditions for that through its records, seeking their best behavior for evolution in relation to the characteristics of the environment to avoid possible genetic regressions in the process, but for this, she always kept his best result for its evolutionary line and kept the other information to be used later when was awaiting new combinations, which certainly occurred, enabling a gradual increase in the complexity of its structure and may be considered as a true software of life, where the slowness in a paradox contributed to greater reliability and efficiency of their records, because they were billions of years accumulating information, until it could create the first basic structural unit of all living beings, represented by the cell, the result of countless chemical records gathered by its joint genetic

In the cell there are two types of nucleic acids, and the DNA that stores genetic information, RNA, responsible for the smooth running of the components of the cell. The DNA molecule is composed of a set of genes, which holds within itself, the chemical code that orients cells in the task of manufacturing proteins, substances that define individual characteristics of all living beings, is located in the chromosomes and each animal or plant species, has a fixed number of chromosomes. Its shape is so extraordinary as unmistakable as it resembles a spiral staircase, the allowing you to perform a unique maneuver in the reproduction process, when the cell divides, the ladder splits in two and each side of the ladder attracts to itself, the missing elements and are sparse in the cell, so that soon form two DNA ladders, perfect replicas of the first, creating in this way, the necessary conditions for the early stages of evolution through the cell group, which became over time, increasingly intense, adding if the experiences of each cell and improving genetic quality of information of their respective genes.

The oldest microorganism that science has knowledge is dated approximately three billion years, because it's very hard to find their traces, which are usually on rocks or sediments. In the early days of evolution, the duration time of a way of life, could vary from seconds to billions of years, depending on its adaptation to the medium, but in these various stages of evolution, the first condition for the development of life a little bit more complex, happened in the water, due to the excellent conditions offered for movement and reproduction, but in with regard to the characteristics of this new being, he did not have a well-defined shape, because this certainly would occur over time, according to their needs of adaptation to means without senses that would also be developed later, obviously without movements for own shares, but with relative ability to Association and evolution, because otherwise, we wouldn't be here today.

Because of the constant attempts of integration between systems, was creating in various environments, beings with some autonomy and minor differences caused many times by various characteristics that the environment offered and to interfere directly in final genetic quality achieved. Over time, the development of the first kind of sense it happened and it was a kind of perception caused by thermal variations, producing small structural changes, making him closer to or away from certain could be propagated originally from sunlight or any heat source.

In this period were occurring several genetic mutations that resulted in a series of new sensory abilities and moves, causing therefore a greater perspective of life beings and also an increase in their differences, which have begun to be more pronounced due to multiple environments offered by water, along with the genetic variety available for possible combinations in each region, which occur through contacts often caused by simple movement of water, enabling the emergence of various species, some of which began to protrude in its forms, facilitating their proliferation on the environment and as a result of natural selection of species, only a few forms of each species prevailed in each region, usually two. Tubular sponges, the braquiópodes and the trilópides appear as the precursor species of animal life on the planet.

The interesting thing in all these steps and that impresses a lot, is their dynamics as a whole, always seeking a better result and preventing any retrogression in the integration of life with the environment, marked mainly by two main links and little dependent, being one on Earth and another at sea. The evolution at sea can be considered as fantastic, because all of our basic features of coordination, senses, memory and movement were developed at sea. One of the most primitive predatory species, appear the Nautílus, which had as its favorite prey Trilópides. In a past remotíssimo on the planet, there was a period in which the evolution of the sea was way ahead of what was happening on Earth, which adopted an evolutionary line different mainly because of the great difficulty of offset within this environment.

However is characteristic of the evolution of life from its beginnings, adapt the peculiarities of the circumstances, and in this line, initially the plants from the sea, grew always intertwined, until certain situations in the environment caused the breakup of these interconnections, stimulating nature to develop simple ways for an independent evolution, allowing the gradual estrangement of these plant systems, which have been acquiring different characteristicsas as consequence of exposure to various environments, enabling the formation of lush plant walls crowned by the tops of trees, only possible manifestations first, due to the gradual vascularization that occurred in the plants, allowing the sending of water through its entire structure.

This system is based on the surface tension of water, property that their molecules have to be strongly attached to each other. This happens because the hydrogen atoms bind to oxygen atoms in any molecule, irrespective of the material, a process known by science as capillary action, i.e. a molecule of water if soil adheres the roots loaded with oxygen, where she encounters another molecule, if linking to successive form, forming a long chain to the top of the tree.

Worth pointing out, that the beginning of this mutation of sea creatures, seeking the life out of the sea, began due to small inroads that some species were in the atmosphere and that was gradually stimulating nature to seek an integration with this new environment, which took place at various points on the planet, with several species, sometimes with well differentiated physical characteristics and when they were out of the water were several different ecosystems, often requiring the development of new skills for the guarantee of their survival, causing them several physical transformations and enabling the creation of several species of animals we know today and they are somehow descended from those first species that if dragged to Earth, about three hundred and seventy million years, finding a set of life formed by plants and insects.

An interesting fact in the first ecological niches, refers to interactions between insects and plants, which became so systematic and intertwined, often confusing on a single system, a real symbiosis and perhaps for this reason, the evolution of insects has been much less optimized, which is easily noticed nowadays, considering that they always follow a pattern totally generic, as well as exemplified by its more evolved specimens, such as ants, bees and termites, which are basically wandering cells always commanded from a distance by a kind of social hormone developed millions of years ago, the result of countless victories and failures of primitive insects, males and females, when were willing to live together.

The livelihood has always been the priority of all living things and obviously when the first arrived on Earth, their respective instincts directed him to integrate with the environment, seeking to maintain their survival, causing a phenomenon that so far had occurred with very dimly, represented by the interaction of systems developed on land and at sea, but now with a much broader optimization in demonstrating the complexity of nature, which from the beginning has as its basic characteristic, the transformation, always aiming to enhance your systems along the Middle, through casual attempts and continuous tests, seeking through a dynamic biased, the favoring of life as a whole, creating and transforming according to your needs more imminent, developing extraordinary vital and systematic that are an integral part of our lives, such as, the immune system, which protects the body from possible antigens, skin considered the most versatile organ and heart, which is a kind of hydraulic pump, capable of maintaining constant circulation and that came about five hundred and seventy million years in annelids or earthworms and developed as a result of the sophistication of life on the planet, becoming required a system to take substances for all nutrients organic structure. He is considered the seat of feelings and thus earned a place of honor in the language and literature produced for thousands of years, this because it changes his rhythm on what you love or hate yourself, truly setting the rhythm of life, which was already in its early days very fragile in his set, forcing the nature to develop defensive systems, mainly due to some bacteria that throughout evolution have learned through experience to camouflage and began to pass unnoticed by the body's defenses, which no longer had the same tune in its activities by stimulating the gene pool to develop the B lymphocytes, which has the function to identify these antigens through the proteins, so they don't contaminate other cells of the body.

These proteins are the so-called antibodies, that has as main function to mark these invaders and call the attention of defenses to destroy it, however if all human skin cells are identical, the same is not true with B lymphocytes, which makes sense, after all need to specialize in the production of antibodies of various sizes and shapes to fit like puzzle pieces in a multitude of enemies which one of the hundreds of billions of B lymphocytes in the body, there are about a million different types of antibodies.

In the course of an infection, some B cells acquire what scientists call of memory, the property that allows them to monitor in detail the characteristics of the attacker, so that in a second time, would B cells specialized for this type of attack and able to act more quickly than in the previous attack, but when the lymphocyte B finds himself face to face with your Antigen, does not set the fire immediately, he expects first antibodies the attack order given by a substance called interleukin, which is sent by the T cell, when you realize the presence of the antigen, ordering the attack, and controlling in this way, all the action of our immune system.

Is nature creating its evasive, through a selective interference, always very strict and objective in their quest for the best results, well exemplified by this fantastic protection cell called B lymphocyte, developed 400 million years. Within this line of protection, she also created the skin, which acts like an armored shell, leaving out the Sun's rays and all likely enemy, using mainly in its versatility to maintain ideal body temperature, keeping the heat on cold days and cold on hot days, reserving water to be used whenever necessary, controlling blood pressure, producing vitamins, eliminating toxic substances, capturing a diversity of environmental information and transmitting others, gives the body its contour and relief, however, few actually know what it's like to be in her own skin, that although it is a very thin, having literally gear a maximum thickness 2 mm, hosts a number of structures such as nerves, glands and muscles, each with quite specific functions and which have characteristics somewhat different in its various points, and can form the thin to eyelid breezes, avoid the wear on the feet, creating calluses or protection on the joints, such as knees and elbows, be folded to allow flexibility or still on the fingers to grab better, possess grooves, anyway she fits the part of the body that is. She has a very short life cycle, approximately twenty-one days, having a curious fact in your evolution, your nails are nothing more than modified epithelial cells (skin), whose keratin maxes out of rigor, however there are some thousands of years without no special function, they are only remnants of primitive times hard in that if needed to fight tooth and nail for survival. "The fascinating history of evolution is long and full of curiosities, the way to see them, depends only on the eyes of each observer." The Scorpions are among the oldest animals, at least 500 million years, represented by a type of aquatic Scorpion that lived in the Paleozoic era.

They have been through successive planetary disasters, glacial periods, intense volcanism, flooding of entire continents, but have gone through everything gracefully and are there, beautiful and fagueiros, giving the name to a constellation of the milky way and even revered in some branches of mysticism, for example, astrology. No wonder that many scientists have come to believe that in the event of a nuclear catastrophe, devastating enough to derail more sophisticated Life forms, the Scorpions would be one of the survivors and beings would continue to wander the Earth, calcined in a strange paradox, seen as the symbol of death, would be the only representatives of life on the planet. Almost everything about them is different, the mode of attack victims, to eat, to reproduce themselves, he is fantastically unique in its characteristics, being the amazing virulence of its bite, just one of the many points of interest acquired by this animal.

Males do not have penis, so the sexual encounter of the male and female occurs so unconventional, which begins with a maneuver that intertwines its talons in the female by dragging it to what you can consider a long walk, because the path is full of comes and goes and extends over a few dozen metro, when suddenly the lower part of the abdomen of the male out two tiny rods that stick on the ground in the vertical position, then he pulls the companion, by sliding of belly on them and right now happens to fertilization, because the apex of each swab is a small reservoir of semen that comes into contact with the genital opening of the female. Little is known about the internal anatomy of its prehistoric, the fossilized remains of those animals, show only gigantic outlines of bodies that measured almost 1 meter in length, having the oldest of all been found on the island of Gotland, Swedish territory, right in the middle of the Baltic Sea, but only a few fragments were recovered.

The largest Scorpion that exists today is Pandinus imperator, who lives in Equatorial Africa, measures about 20 cm and shines at night like an unsettling reminder of their giant ancestors.

He is very exotic, the Greeks gave it the name of hippocampus and everything in it is curious, it's the female who takes the initiative of dating, but he's the one who gives birth to the puppies, no mouth but eats a lot, your eyes moving independently in the eye sockets, your skin changes color depending on the circumstances, your tail is prehensile and incubates its young in a pouch waist belt.

Its shape is an elongated external skeletons coated stomach, propping up internally the muscles and other body components. Is a fish, but the only trace that the denounce, is the presence of tiny and almost transparent dorsal fins. He seems to live cooped up inside a real arthropod exoskeleton and the skeleton so the skin, gives the impression of being always starving.

Sua boca está realmente no focinho, a ponto de, um canal estreito e comprido, onde o alimento é sugado em direção ao estômago de partida. Estes habitantes curiosos dos mares tropicais estão espalhados por todo o mundo em cerca de cinqüenta espécies e devido à sua facilidade de mudança de cores, se escondia com perfeição entre os vários tipos de algas marinhas que constituem o seu ambiente favorito, mas são freqüentemente encontrados ao longo da costa em águas rasas no recife de coral. His sex life is the highest point of all its oddities, the female begins to woo the male, with a discreet touch of waist and then throws herself at boulder caresses, linking it with the tail, when you deposit a kind of gelatinous mass (ova) in her handbag, causing the debatable Stallion, the growth of a respectable little belly, that after fifty days, will give birth, about three hundred tiny "Colts raring", which will be inside your handbag for a few weeks, until they leave.

For the ancient Greeks, this animal represented a massive poison, since soaked with wine, but a powerful antidote, when swallowed with honey and vinegar, but all this is superstition.

Within this context, this unusual animal, could not be framed in compendiums of Zoology with a fish any, so the sea horse, became scientifically a respectable Gasteosteiforme.

Few things are so perfect in the world, as they, the masterpieces that nature took hundreds of thousands of years to produce. The varieties of their shapes, colors and combinations of as wide, seem endless, making them one of the rarest and most beautiful spectacles of nature, especially when they mingle with the clear water and the Sun's rays, causing them a singular effect of splendor, filling our eyes or especially when they are grouped as a tourmaline bicolor or small Topaz crystallized on quartz, really enchant by the sheer beauty that express, are natural components of the Earth's crust and feature a well-defined chemical structure. This structure represents the way you arrange the atoms of different elements that form a mineral and for this, she has decisive influence on the determination of physical and chemical properties of each one of them.

Good examples are the graffiti and the diamonds, both constructed of pure carbon, however have different crystalline structures. The graphite is commonly found in rocks that are formed at the top of the crust, but the diamond is much rarer, it forms deep in the crust, in very special volcanic rocks, where pressure and temperature are very high and therefore its carbon atoms constitute a much more compact structure in the form of four-sided pyramid different water, Navy and Emerald that form hexagonal prisms, already Topaz crystallizes in the form of Prism-shaped basis. Within the esoteric, a chain or Crystal Amethyst ring, can mean more than mere decorations. Insiders say that a quartz inside an aquarium makes the fish more clean and shiny or near a vase, would make the plants grow faster. Some supporters of, acupressure, Crystal therapy believe that everything can be explained by the transmission of energy contained in the crystals.

For those who appreciate their good quality of crystallization and the perfection of its forms, are very well represented in the Museum of natural history in Paris, in one of the most amazing and important collections of crystals in the world. "Despite the crystals being inanimate matter, they also have a history of evolution as life." There are approximately six hundred and fifty million years ago, the archaic era existed, formed by the arqueozóico period, having the life at sea started in this period.

Today, older species are represented by the jellyfish and the vermin of the worm type, discovered in South Australia. A hundred million years later we had the Paleozoic era, marked mainly by the fantastic proliferation of marine species, though all are extinct, the marine fauna account today with some descendants around in its shape of beings from that era.

The Mesozoic era began about two hundred million years, while the appearance of the first birds that mixed the plumages, with traces of reptiles, and may be regarded as a great evolution, because besides being a magnificent thermal insulator, obviously constituted a fundamental aerodynamic element to them. The current era is called Cenozoic, when appeared the first warm-blooded animals, showing much faster biochemical reactions, making their response to external stimuli.

Ten million years ago, our ancestors were just species of monkeys that live in tropical forests that covered Africa, however during the process of evolution, the planet itself began to get colder and drier, the African forests were dwindling and the steppes expanded, however, rather than become extinct or continue clinging to what was left of the forests, went to find food in the savannas, and according to some anthropologists, this quest became more effective as soon as they got up and began to walk on two legs, being being that these creatures evolved bipedal hominid, our oldest ancestor.

After this, over a million years, the African climate was moving from cool and dry to hot and humid, until, about three million years ago, there was a radical change, a great dry cold front initiated the latest series of ice ages of the planet, forcing our ancestors to walk across the globe, winning the planet's climatic vagaries, evolving its physical structure, enabling the emergence of higher primates, anthropoids and ancestor of homo sapiens, marked mainly by increasing its cranial capacity.

When homo erectus first entered in a cave with a torch in hand, must have been a real glory, for surely the flame lit and warmed their environment, made the fierce beasts, gave rise to the barbecue. However, half a million years later, despite the fire have already bustling reactors and have helped like no other event, the construction of civilization, the man doesn't know you yet as it should and I just learned a way to produce it there are only nine thousand years, therefore already in the Neolithic period.

Probably the first fire maker, must have noticed a spark produced by friction between two stones and to reproduce the phenomenon, must have experimented with different types of rocks, until they decide for the best, as the Flint and pyrites, found in archaeological excavations. The natives of the Islands near the Andoman India and some tribes of Pygmies of the Congo in Africa, for example, has ever managed to light a fire without from hot before, but ended up learning from other peoples.

Nowadays, scientists are still trying to learn the innermost secrets of the fire, after all he is a phenomenon that in addition to the chemical reactions provided happens too physical processes such as mass and energy transport, diffusion of heat and radiation. The color of the flame is a combination of temperature with the burned out element and provides a way to fathom what is going on in apparent confusion of the flames.

During your Burns, she acquires a characteristic color according to the burnt substance, for example, copper causes a greenish flame, sodium emits a yellow flame. This happens because the heat excites the atoms of the substance that is burning, causing the emission of light. Today Science uses these luminous radiation spectrums, to set even the composition of stars. "With the discovery of fire by primitive and the beginning of language, which occurred about 50,000 years ago, defined the first steps towards civilization."

The first great empire known in the world was created about four thousand and three hundred years by the dreaded King Sargon of Akkad, who grouped under his command prosperous independent city-States and villages of Mesopotamia, the so-called cradle of civilization, located at the time in a fertile valley, between the Tigris and Euphrates rivers, which occupied parts of what are today the Iraq, the Syria and southern Turkey.

The ancient Egyptians were obsessed with the afterlife and their rulers sought immortality by erecting massive stone buildings, establishing in this way, the foundations of the first great nation-State in the world. When the subjects of Pharaoh Djoser contemplated for the first time its gigantic pyramid-ladder, must have blur of amazement. The colossal monument, in the dusty plateau of Saqqara, 15 kilometers south of the Sphinx and the pyramids of Gizeh, which was designed to inspire awe and impress the Egyptians with the strength of its ruler almost divine.

Being at the time the largest and most beautiful monument ever erected by a monarch, the largest building in the world. Just like the pyramid, ancient Egypt seems to have grown out of nothing, because a few generations before the reign of Djoser, civilization along the Nile River was nothing but a group of nomos, small provinces with Government and gods themselves. Scholars only have vague idea of the forces that caused these nomos, always at war, becoming, along with the Sumerians in Mesopotamia, the earliest civilization of the age, the ancient Egyptian Empire.

The Renaissance period was certainly one of the most fruitful of the entire history of civilization, being first and foremost a powerful artistic and literary movement with major repercussions on philosophy, science, in political thought, in fashion and in customs. The multiplication of universities and the discovery of the press by Gutenberg, which replaced the laborious activity of the medieval copyist that reproduced by hand, the precious manuscripts of the era, allowed a wide diffusion of knowledge, opening way for modernity.

During the 15th and 16th centuries, the main centres of the Renaissance were the cities of Florence, Venice and Rome, besides the duchies of Mántua, Urbino and Ferrara. The ideal of the Renaissance man was marked by a belief in an unlimited capacity of human creation. This idea was embodied primarily by Leonardo da Vinci.

With his paintings and engineering projects, which made him famous and courted by powerful at the time, but only long after, the world would come to know the secret side of this superlative genius, painter, sculptor, musician, architect, military, civil engineer and inventor extraordinaire, and may be considered the Supreme version of man of the seven instruments that performed the marriage of art and science. Born in the small town of Vinci, near Florence, would be considered in a short time, the greatest painter of his day, protected and adored in some of the major European courtsci.

Was a scientist of extraordinary talent, able to join into a single formula, the fall of an Apple and the motions of the planets, was born premature and weak, but he lived to be eighty-four years, leaving us as his greatest contribution to the theory of Universal gravitation. That through his theory of relativity made a revolution in human thought and today is considered a synonym of science, however up to three years, didn't speak a single word, the nine had still so difficult to express themselves, that his parents feared he could be mentally retarded and at school a teacher prophesied prophesied that he would be nothing in life, but at the age of twenty-six years, however, would publish his special theory of relativity, one of the most extraordinary revolutions in the history of ideas, only reaching a size comparable to the Greek Aristotle and the English physicist Isaac Newton. Considering that his theory would be the founder of Contemporary Physics milestone, with deep repercussions in other branches of science, after it, ideas such as space, time, mass and energy would no longer be the same. "Nature often overcomes and toast with these real gems, synonyms of the most absolute expression of human intelligence, but would never be the same.

""Nature often overcomes and offer us these real gems, synonyms of the most absolute expression of human intelligence, but never forgetting the light sensitivity of the matter. "Green plants began to produce molecular oxygen from the atmosphere in the Carboniferous period, but the code for the appearance of flowers, only really happened in the Cretaceous period, about 100 million years. Their leaves earliest fossils date from twenty-five million years and were found in various parts of Europe and the Middle East."

She always had a story of much symbolism for humanity and shows us that the Romans were passionate about roses, the Greeks by violet and through them, the flowers have become symbols by means of which people express their feelings, each of which has a special meaning, the Tulips are friendship and sympathy, the lilies symbolize the wish of luck and Rosa considered the classic flower of love expressed suffering and passion.

Na Idade Média, dando um buquê de violetas para alguém era símbolo de um amor secreto e, portanto, os mais tímidos se valem deles para expressar sua paixão, enquanto os cavaleiros que estavam no mundo, desfazendo a justiça desleixado e defender, usavam rosas em seus gorros de veludo, sempre que eles estavam esperando os favores de uma donzela indiferente ao seu amor.

At this time the healing power of flowers reached its apogee, when it was believed that the roses were a remedy that cured all.

There were even those who thought they avoided epidemics, as happened with a French King, who in charge of a crusade to the Holy Land, arrived in Tunis in Northern Africa and roses sprinkled on people, thinking that this would be helping to combat the plague that ravaged the medieval world

The modern roses little resemble their ancestors born in Persian fields, millions of years ago and then perpetuated for the rest of the world, as if the simple farmed species, and beloved by our ancestors had pale color, fragrance, and only five petals, the current generation blooms with a hundred petals, displays the most varied hues and perfumes the air with its striking aroma. In addition to expressing an exuberance that Flora, Greek goddess of flowers, would be proud, but it's unique fruit from the hands of the man, who in a rare moment of inspiration, interfered in nature, not to destroy, but to enhance the rose, intensifying their beauty through a genetic process that may take centuries to develop naturally.

Is characteristic of the evolution, seeking always to take advantage of the genetic experiments of other bodies, through any form of interaction possible that might happen within the environment, always seeking some way in the future, test these new genetic possibilities, this is the genetic improvement cycle and occurs with all beings on the planet, animals or vegetables. Of course, to further understand mutation more apparent, can take hundreds of years, and by this I perceive clearly, because our life cycle here is very brief, what are eighty years within a process so long like the evolution.

According to science, the natural duration of life, should be between sixty and one hundred and thirty years, because it is the time that the cells of the body leading to aging. To be young in the concept of genetics, is to have the cells working in harmony as an orchestra, however she begins to go sour in the course of life, in reason of the loss of atoms that make up the molecule of nucleic acid causing aging.

Today with genetic manipulation, the possibilities in the quest for longevity are many, after all is our gene pool or genome that controls everything. Our specifically, consists of twenty-three pairs of chromosomes, which symbolically can consider as an encyclopedia. Each chromosome is a book containing thousands of pages, where are all our genetic information. The gene would be the chapter of this book, with the function to synthesize proteins, that define the characteristics of beings

The word would be the triplet and would specify the type of amino acid to be used and the nucleotide would be the letter. In the genome there are only four letters or basic substance that are called adenine, cytosine, guanine and thymine and who wrote the entire evolutionary history of beings, and the thymine only pair with adenine and cytosine paired with guanine form only..

A sequence of three bases forms a Word, but although there are sixty-four words that are resulting from the combination of four letters in groups of three, only twenty are used, each of which corresponds to a type of amino acid, the forty-four remaining have no clear function, can be used only to indicate that genes have provided the complete sequence of a protein. Today with the mapping of the human genome, science seeks a better understanding of how each molecule travels inside the cell, how genes are switched on and off.

It's a task no doubt quite pretentious, but this continuing with astonishing rapidity. The Wisteria is the simplest raw material of proteins and emerged second laboratory simulations, in the ocean, from the combination of hydrogen cyanide, ammonia and methane molecules fragments caused by the ultraviolet light from Sun.

An important observation is that in addition to life cycle, there is also the genetic cycle of the species, whose duration is directly proportional to its ability to bring new genetic variants for their respective genomes, i.e., if the quality of genetic perception is diminished by a very long period, which can range from hundreds to thousands of years, will cause a gradual locking in reproductive capacity, making the evolutionary line, abandon this nature is what would be an answer, regarding the purpose of beings and of their respective genomes without good quality genetic interaction, would be a lifecycle obsolete, and that's what happened to the dinosaurs. They ruled the Earth for millions of years, were giants vegetarians or could be small carnivores, cold-blooded like reptiles or warm-blooded like mammals, violent or somewhat clumsy, lived in the midst of lakes, rivers and forests, under a mild climate and landscapes formed by Prairies, mountains, deserts, swamps, rivers, plants like pines and ferns.

The water of the oceans was tempered, being inhabited by sharks and other fish. Absolute Lords of the planet for millions of years, there has been to date species that would entrance him so much and had so much curiosity. What is known about them is very little, nor is it possible to confidently assert that fed or how they became the dominant species on Earth, after all reigned throughout the Mesozoic era. Appeared in the Triassic period, about two hundred million years.

Crossed all the Jurassic and the Cretaceous, when became extinct about 65 million years ago. The first discoveries of teeth and bones, only occurred in the late nineteenth century and were found casually in southern England.

Today it is known that they had some sensitivity, because in herds, the adults always were protector of younger. "This restlessness of nature, in his constant search for new, was also transmitted to man and motivates your curiosity in General. Today the man always tries to relate the technical aspects of life, with its social realities, a good example, is the quest for artificial intelligence. "Create a computer able to invent stories, learn from their mistakes and have emotions as humans, this can be called artificial intelligence.

"Create a computer able to invent stories, learn from their mistakes and have emotions as humans, this can be called artificial intelligence. The machines of last generation, already has eyes, ears, touch and are able to talk and make some decisions, demonstrating the real revolution that's going on in the world of computers, at a level as fundamental as those of the neurons, the cells of the central nervous system and that some scientists call of neurocomputadores technology"

These machines try to imitate our neurons and are already very evolved in the field of processing and recognition of signs, but are a long way from playing more complex functions of our brains, because their units are homogeneous, while the brain has hundreds of different types of neurons and a number of different areas of expertise, a vision is another of language, etc..

For a computer is very difficult to accurately interpret what they see, for example, the complexity of the human eye is formed by more than one hundred millions of receptor cells that process the information of light stimuli to the brain, up to this point everything well, computers translate to numbers, in this case, each point of light the image you see, but the mystery is the manner in which the human brain interprets the information from the eye in this aspect, the technology is still far from a satisfactory result.

The plethora of electrochemical messages that our billions of neurons Exchange allows us to have sensations, movements, motor coordination, memory, creativity, sound judgment and abstraction. It is possible that in the future there are machines capable of self program, coming to a near-infinite intelligence, developing even some kind of consciousness, which allows central monitor their own States, in a pseudo immortality residing in Silicon cells, however they never suffer the vicissitudes of death, for example, altering their parameters of analysis in the field of emotions and for that, whatever the notion of itself that can engender, she will be far from human nature, due to its precarious expression in the field of emotional intelligence, which existed for about a million years and be smart in our day, emotional means know interpret its forms of emotions set your goals always considering the time and energy available to devote to a proper way to several plans of life, never disregarding the social effect of their decisions that becomes increasingly increasingly diffuse, when you enter this wonderful world of modern technology, we are joining an amazing Decade for communication.

Probably when she's finished, all forms of communication in the world, whether voice, data or images, will be transmitted through a single optical fiber, in just a few seconds, our known transistors can be made of just one atom or our processors are no longer than a molecule. It's amazing the possibilities of nanotechnology which will allow in a few years the production of tiny robots or even the big breakthrough in the biological area, which has made possible the development of new software for use on computers in computers increasingly fast and efficient.

Today we are much closer to the quantum computer, capable of overcoming all the human imagination has conceived, because the limit of current technology in the atom, perhaps you can go beyond, into the nucleus of the atom, but it would be much more complicated, although the future of nanotechnology be inside the nucleus of the atom, yet we know very little, however the science think of multiple ways and also acts in optical computing, biological, analog and digital.

Quantum computing is something extremely new and is based on the manipulation of electrons one by one or in groups, joining them or separating them, but even if this at the beginning of this path, which makes it a fascinating subject mainly for its possibilities.

In the case of biological computing, it is based on the DNA molecules and proteins, which added to the advancement of physics, chemistry and biology, will be possible to create new concepts of computing. Through the meeting of science and physical chemical biological techniques, scientists are learning new ways in the synthesis process, only possible through nanotechnology and are seeking too, understand the workings of the brain of primitive animals, such as the slug and snail, with the purpose of transferring part of such knowledge for the field of artificial intelligence and space technology, favoring future planetary explorations, in the search for extraterrestrial life and diverse knowledge.

There are many people who swear to have seen a flying saucer, science doesn't take this really seriously, however there are a few appearances that have never been well explained. Many of these visions psychoanalysis explains as being hallucinations caused by collective anxieties, which occur in periods of crises, it would be, so a modern version of the vision of Saints and demons so common in the middle ages. According to this interpretation, the space-age man expects to be safe from their everyday problems, not for angels like they used to, but by extraterrestrial beings.

Within this context, it is understandable why seeing a u.f.o. is easy, the hard part is to get someone to believe. The mystery of the Ufo emerged soon after the second world war, when an American pilot claimed to have been chased by a squadron of ships in the shape of a saucer, not taking too long to the tabloid press design in the popular imagination, which were even Martians invading the Earth.

In the future it is likely some kind of contact, which may be accidental or deliberate through the tracing of signals in space, after all it is possible that in this vast universe, where the Earth is a point less than negligible, there are many worlds on which some form of intelligent life has flourished, although nothing to indicate up to now, that some of these worlds is one of the eight planets in this solar system but that exist at some point these one hundred and twenty-five billion galaxies in the universe, the laws of probability, become more than plausible, however if your form of evolution follow the same principle of our planet, all our differences will depend on just two basic factors, when and where this extraterrestrial life began, with the "where will define its physical characteristics and the" when "their degree of development. In this fascinating journey of knowledge, scientists will gradually discovering things really extraordinary about the universe. As the quasars, which are celestial bodies that Flash the billions of light years from Earth, but were discovered only a few decades.

It is true that glow in the space intensely, there are about fifteen billion years, since the origins of the universe, featuring a small and almost insignificant point of bluish light in the dark sky, which can be seen in the constellation of Sagittarius, having been almost witnesses of the Big Bang. Are the most distant bodies that we can identify and take part, so the role of the fascinating puzzles that challenge man's curiosity.

Now we're entering a theoretical world, very different from the macroscopic reality with which we live every day, after all we are talking about the internal structure of protons and seeking the last frontier of reality, the basic bricks of the entire universe would be formed, but first let's review some concepts about the atom, which represents the smallest portion of matter. The temperatures of the early universe, matter was disintegrated into a plethora of elementary particles, which were gradually cooling estimated at more than three hundred thousand years, until the appearance of the first atoms.

They are formed by protons, electrons, neutrons and other quantum particles called quarks, leptons and gauge bosons, being that protons and neutrons form the nucleus of the atom and electron is orbiting around this core. Protons have positive charge, the negatively charged electrons and neutrons are devoid of load, making it possible to verify that most of the volume of the nucleus of an atom is occupied by empty and he keeps himself thanks to strong nuclear force, if she no longer exists, the core would explode due to electromagnetic repulsion between protons.

The strong nuclear force acts only at very small distances on the order of two or three times the diameter of the particle itself, however if they get closer can become repulsive, preventing the nucleus of atoms burst due to electric repulsion.

The magnetic, gravitational forces, strong nuclear and weak, are the four known forces in the universe and all act in the nucleus of an atom, which under the strong nuclear force, all existing matter in it, appears extremely compressed. Through quantum mechanics, know, that in any type of radiation, for example, light may be emitted, transmitted, and absorbed in discrete amounts of energy, meaning that the flow of energy is formed by a number of small individual packets called quanta (plural of quantum), and the energy of each quantum is equal their radiation frequency multiplied by a constant value called the Plank constant.

This formula helped explain the photoelectric effect via the photons, which represents the light as a stream of particles and not as a continuous wave, and can be bent by the force of a certain amount of mass, like a star, however, as the speed of light cannot be changed, is the time that adapts to its curvature, keeping it constant, even in extreme situations, such as a black hole, stellar phenomenon that occupies the center of our Galaxy and whose gravitational pull is so intense that not even light escapes, acquiring a black shape.

The theory of relativity says that the time is not something essentially different from space, so in addition to the three known dimensions, length, width and height, the universe has a fourth dimension called of space-time.

This dimension space-time is flexible because it contains a high concentration of mass, for example, all the planets in the solar system, are held in their orbits due to the curving of spacetime produced by the huge mass of the Sun. Our current technology allows us to achieve a speed around 40,000 miles per hour, however we are far from the speed of light, which is around three hundred thousand kilometers per second, but travel at this speed would be impossible, because all matter subject to this speed transforms instantly into energy, however, we can get very close to it, but for this, it will be necessary to develop a propulsion system powerful enough to propel a spacecraft with very high speeds.

The first option would be a spacecraft powered using the photon, this energy would be resulting from the fusion of matter and antimatter, another option would be to use hydrogen, most common element in the universe or even a rocket to ions, which would have its energy from the disintegration of ions, it doesn't matter, probably in the next few centuries, man will develop technologies for approaching this speed but our reality within the spacetime is much more complex than simply reaching the speed of light, for example whereas the nearest galaxy which is Andromeda, has distance of about two million light-years from Earth, we're never going to know any Galaxy in this vast universe. Of course, this statement would only be valid if it considers that the space is something absolute, and thus the distance between the Earth and the Andromeda Galaxy have nothing to do with the speed of the rocket, even traveling at the speed of light an earthling would take 2 million years to reach

However the wonderful in theory of relativity demonstrates that is not so, because the higher the speed, the greater the relationship between space and time, meaning that the speed is nothing more than certain space traversed in a given time, making it clear that the distance is relative as all other parameters that surround us, after all our universe does not represent everything Since it is a result of manifestation of a phenomenon, who obviously couldn't have come out of nowhere, being more logical if you think about the existence of another universe, another universe, coexisting with ours.

When working inside of relativity, the journey of an astronaut in a spacecraft, whose speed is ninety-nine percent of the speed of light, would spend approximately twenty-eight thousand years until Andromeda, fantastic time savings compared to previous hypothesis, but if the spacecraft was traveling to ninety-nine percent of the speed of light added to decimals after the comma would lead briefly but four years However, this contraction of time, can only be observed from extremely high speeds.

Between the departure and the return of the spacecraft would have been spent just over eight years for the astronaut, but for those who stayed on Earth, would have been fantastic three million years, so we're going to have to get another way if we are to meet again in Andromeda and our time.

It is believed that in the universe there are no really empty points, it is possible that all he is expanding within a type static ghost mass, which lies on the boundary between space-time and the absolute, so named because it would be made up of tiny particles, called undefined particle, which would be far beyond the quarks, managing to cross the field of imperceptibly, remembering very generic neutrinos, which are generated by the explosion of a supernova and well known for our science, but they would be much less dependent on our physical laws, their structural basis would be another. When you think of subatomic structure, the space between the particles becomes very important, or rather, the supposed emptiness that exists between them.

The basic difference between the emptiness of the nucleus of an atom and the void between planets, is the emptiness of the atom this much closer to breaking the barrier of space-time, being that this closeness is directly proportional to the concentration of matter, i.e. If there was a concentration close to a black hole, this barrier space-time would be brokenness supposedly happens to black holes, causing two orthogonal paths of expansion of the universe, being a caused by Big Bang and another caused by black holes, because due to a strong distortion that happens close to your core, much of the area attracted to him, was supposed to be released in the form of continuous jets out of the curvature of space-time that defines our universe and due to an abrupt change in speed, would reach fantastic proportions of energy, and can form even galaxies.

Today science can say with a certain degree of certainty that protons and neutrons have an internal structure, represented by Quark, which theoretically holds aggregates by a force so powerful, it would be virtually impossible to pry them out of there, being caused by an intense interaction, resulting from an exchange of particles between particles, i.e. protons and neutrons would remain so why would they be atomic nucleus in agglutinates constantly exchanging particles to each other.

The more you research the intimacy of matter, more surprises appear and one of the most used paths for this internal inspection, are the particles of subatomic dimensions, issued by Non radioactive substances such as uranium that emits different types of radiation, which received the name of Alpha, beta and gamma.

The alpha rays are positively charged particles, the rays are high energy electrons betas and gamma rays electromagnetic radiation similar to simple light, but much smaller wavelength. When there is a very dense concentration of matter can be thought that the particles inside the nucleus are immobilized, which is not true, because the particles when confined to a small region of space tend to a frenetic movement, protons, for example, when confined to reach the dizzying speed of sixty-four thousand kilometers per second. There are situations within the nucleus of the atom, which bills itself as if reality need to respond the same way, no matter how the question was asked and concepts such as particles and waves mix on one, when it seeks to understand the subatomic world.

The duality of matter that behaves as a particle or as a wave, creates situations unimaginable to our common sense, facts that seem paradoxical, because we compare with our macroscopic experience and who are very far from reality unless the world microscopic energy, for example, the wave nature of matter, is not restricted only to the subatomic world, which in principle, is not out of place to say that all the bodies in the universe have an associated wave, this applies to all living beings, as for the planets, stars, galaxies, and this wavelength is so small that escapes a more accurate detection.

When we think seeking situations before the Big Bang, everything becomes extremely enigmatic, getting impossible not to move in a more daring in the realm of imagination, to get some answers and in this line, we can consider that before the Big Bang, there was just some sort of estateticismo, which in a paradox, what happened could be considered, a tiny demonstration any, causing a growing imbalance of this at some point of infinity.

If we consider the observation distance symbolically in a trillion years in this context then, the observer would have seen our universe with its fifteen billion years, as nothing more than a simple point of light when seen from this distance, which could also be looking at perhaps several other points of lights, meaning also several other universes, perhaps even with some similar to our, which were also expanding and up who knows if finding in a distant future, everything is possible, but to the extent that this initial imbalance increases proportionally to release also started some kind of energy, causing a this expansive phenomenon, of which the higher was the concentrated energy, the greater would be the three-dimensional space also designed. In this context, it is relevant that before the space, time and matter, there was most likely there was just this energy and space-time only started to exist, the instant matter was acquiring a three-dimensional State, defining the space and time that came into existence with the expansion of the universe.

On a psychological level, our conscious experience of time depends necessarily the way we perceive, because the line between the past and the future is extremely tenuous and who moves it is our conscience, which reacts very slowly the sequence of facts. The present is a moment so infinitely short, who becomes delusional to our reality and therefore, we can say that there is no time to be valid universally, because there isn't any now that is equal from one point to another of the universe, meaning that the concept of this is a matter purely of reference to the observer, depending on only their State of motion.

The time is closely related to the space, being both physical phenomena and through matter, one cannot exist without the other. It is possible that matter as we know it is made up of superimposed layers of energy, and the power set of each layer would express a set of particles and that would in turn and down until you reach the mass of the proton and the neutron, being that its base would be made up of billions of particles, which pulse independently, but always adding their energy results, however only would interact with the space-time dimension when they were on the move, when static, would become totally imperceptible and independent our physical laws, it would be like if there were no really.

Within these considerations, the matter could be a simple expression of energy that enabled the process of life, through the chemical combination of molecules. Everything is kind of understandable, but in no event, perhaps there is a special State of matter in which the space-time dimension would have little influence on her.

It would be great if in the future this was true because it would open a fantastic range of researches and of possibilities, once unimaginable to our reality, changing all references we have today of space and time, but as we know, science is time-consuming and I haven't even broken the technological barrier of light, this is thinking ahead, a fiction, what is truly right, is that the difficulties of the principles are part of human nature. She comes from a long evolution genetics, which added to a personal trajectory in a physical environment, social and cultural determines our behavior in General.

One of our main features, highlights the central consciousness, which is based on the older regions of the brain, having your action independent of the more complex mental operations, allowing our intellectual efforts interfere directly with the cognitive processes, which lead to manifestation of emotions.

Although the genes have directed all the organic evolution, including our brains, he does not control our mind, in fact, we can define it as a connection between the body and the brain, which is given by a hardware type biochemist, allowing the performance of language, and the ability of reasoning and its expression mainly through emotions, bringing us the surewhich the human creation is not simply the result of a cold, systematic evolution, but that there are a set of values and behaviours pre-set within our lives and and that helps us to evolve, filling of hope about our real future in this whole process. In this great adventure that is life, we expect a daily miracle, but we forget that this miracle is within ourselves, life itself.

We live in a shower of sensations, of which a tiny part attracts our attention and that we seek in religion, art and science, the meaning of life, but the experience teaches us, that not all paths are for all walkers.

This is because some forget help your aurora inside the sunrise and end up living an lack luster existence, which doesn't mean success or money, but simply a balanced life and well lived.

Faced with the undeniable complexity of creation, sometimes the wisdom and knowledge of the people seem ridiculous, but abolish the difficulties of the principles would be inconceivable, since it would imply the negation of human nature itself, our mission is to decipher them, but for this, we must always be open to all opinions, analyzing and filtering all information, through a self perception of values a sort of intimate dialogue, used for the clarification of ideas, because our thoughts cannot be prisoner of values established in our environment, but we must always be looking for new values, which are tuned in a conscious harmony between time and eternity, which is the source of all wisdom. Generally speaking, there are five aspects of human nature, physical, emotional, intellectual, moral and spiritual, being essential a relative balance between them, in order to have a productive and happy life.

Our reality is almost always less dramatic than the vision that we have of her, should always be an optimist alert, renewing our energy all the time, purifying our thoughts with a healthy imagination, trying to keep my mind always vigorous and smooth after the skin inside, all problems are psychological.

For the optimist, what matters is knowing that now, at this very moment, are being born, growing rich and pure opportunities, that the noblest essence of life makes flow in millions of beings, in all parts of the world, because who of us has the privilege of losing the flow, the vibrant abstractions of life which only manifest themselves through positivism.

"All his wisdom lies in the subtleties of information travelling through time, added to the designs of the universe, which weaves transformation immortal intelligence, showing us his ample interests and forcing us through the multiple talents that we use with a certain boldness between the real and the imaginary, which will gradually forming the superb panorama of our science but that is the reason that briefer created it. "

Throughout history, man has suffered greatly because of their little understanding as regards their real origins, seeking answers in different religions and popular superstitions, in an attempt to fill its empty interior and which often turns out to confuse you even more not contributing anything to your spiritual development.

Among the many dogmas and precepts of various religions that have contact, it is very easy to get lost in the decisions and by this the intelligentsia becomes so important to our inner balance, there is great wisdom in various religions and should always be receptive to all information that will help us in our self-knowledge, after all reality is very hard when viewed from the front, making many people sticking in amulets and other symbols, hoping to be able to share your burden of problems, however this way is not the best, as well as keep him from reality, does not solve their problems.

The science through their research, shows us that the mind has direct influence in our immune system, confusing many times these people in their faith, which assigns your improvement of any illness to some sort of amulet, but it actually was the simple result of a positive mental attitude, which changed its immune system, improving his physical condition.

It's consistent, imagine that in the absence of a suitable intellectual training, faith degenerates in bigots.

Our spirit must be in tune with God, as well as our thoughts must be in line with the life in the pursuit of consistent, since this is the only way to make real our existence.

The science can be anything but self-sufficient, because separating the religion of science so they don't fight each other, is only half the story

There are many ways to communicate with God and each must find his and after this, surely the purpose of life will become increasingly clear in your mind. Religion has always been very important to our inner balance and there was a moment in our history, in which became necessary more empiricist, so we could better meet our real intrinsic values and we needed a great reference.

Every year on the twenty-fifth of December we relive the Supreme Christian tradition of Christmas, when we celebrate the birth of Jesus, people exchange gifts, decorate pine trees with lights and colored balls, assemble Nativity scenes depicting the birth of Jesus, the Grotto in Bethlehem, the child in the manger, the shepherds and the three wise men who according to the Bible came from the East bringing gold, frankincense and myrrh, guided by a star in search of King promised, in this way came to Jerusalem, capital of the ancient Kingdom of Israel, where reigned Herod, however there have not found the newborn child and followed to Bethlehem.

The offerings that they presented Jesus were loaded with meanings, Gold symbolized the royalty, the incense to divinity and myrrh was used in burials, which makes suppose that Matthew was alluding to the death of Jesus. In the Biblical sense, the Magi represented Nations, who recognized the boy King promised, are all of oriental origin and they all had to do with the royalty and power, with Melchior the Hebrew means "King of light", Baltazar of Aramaic means "God protect the King's life and Gaspar is one of the three, which had the greatest chance of being really a King, but who was Jesus anyway? The answer is difficult, mainly because the only accounts of his life are the Gospels, written and rewritten decades after his death. The Church accepts only four of these texts as valid, the so-called canonical Gospel attributed to mark, Matthew, Luke and John. At the time Jesus was born the Jews awaited for leader that dispose of the Roman yoke.

For Matthew, Jesus was the Messiah and that his birth occurred in Bethlehem, being greeted by the appearance of the star of David, but as Matthew's account, Jesus descended from David through Joseph and his action went mostly among the poor and marginalized of his time, spending most of his life in the region of Galilee, which housed a population mostly, totally miserable.

There were almost thirty-year gap in the narrative of the Gospels, from birth until the beginning of his preaching, giving room for all kinds of fantasy, until he would have been in Tibet, which is unlikely. Jesus was not a priest, rarely preached in their synagogues, their actions and their thoughts were always among the people. Baptized by John was fasting for forty days in the desert, this isolation, which happened also with Buddha and Mohammed, because at that time it was normal to go through periods of solitude by nature, before an important decision.

Jesus suffered three temptations involving wealth, prestige and power, yet the crowning point in his career, for which converge all evangelical narratives, was his stay in Jerusalem, where he confronted directly with the power Center, being tried, convicted and crucified, although upon his arrival having made it quite clear through his words, that he wasn't coming to lead a military rebellion against Roman rulebut proposing a transformation into another type of structure of society and human mentality.

Climb the first rung with faith.
It is not necessary for you to see the whole staircase.
Just take the first step.
My God, I'd like to thank you for the countless times that you saw me better than I am!

HE WON THE WORLD

Words of faith

Every day, every morning God prepares always something new for us, so stay tuned, don't distract from the presence of the Lord. Raise your eyes to the lot where my help cometh. Hope assumes the wait soon comes the King. The Word will illuminate the narrow path home. The promise I will fail my believe in you Lord, for the most beautiful work that came out of human hands was the writing of the Gospels and nowhere else will you find deeper knowledge of the human being, its contradictions, its weaknesses, most secret movements of your heart ...

Do you want to be happy? Observe the children. They don't regret the past and don't care about the future. They simply leverage the gift with all that it offers. If it hurts, they cry.

If it's good, distribute laughter contagious. Simple as that. Release your feelings and live. God is within you and around you, and not in stone castles or mansions of wood.

Lift a stone and you will find God. Break a piece of wood and He will be there. Who know the meaning of those words would never meet the death ... Sometimes we think we're alone?

We're not, because God is with each of us, supporting us for the fight of your life.

This our time here is an oblique and undulating surface that only memory is able to move, and approximate than it really is, true. In generalised world woes. more don't get discouraged because Jesus Christ won the world .... In our art of living, we must learn from the past, but don't live in it. Everything in life is experiences, you might even say I didn't understand what I said. But you can never say you don't understand how I looked.

The words are full of falsity or art, but the look is the language of the heart, and for this the most beautiful love phrases are said in the silence of a look. Is it fair to think that all those who do not understand a look, either you will understand a long explanation!

The world in which we live depends on how we direct our perception.

Follows a good example ...

Two men looked through the bars, one saw the mud and the other stars. Always seek the horizon and no regrets, since it makes no sense to look back and think: I should have done this or that, I should have been there. It doesn't really matter. Let's invent tomorrow, and stop worrying about the past. What's done is done!. In this line, we should look hard for ourselves before we think about judging others.

The strength of A gaze brings us to maturity and through it we learn to live with less suffering, along with more tranquility, want with more sweetness and especially bring closer look all those extraordinary visions of things that hug with my eyes closed, being very unforgettable.

Jesus is the cornerstone of our fragile lives and the only link with God.

Being that the Eucharist is a celebration in memory of his death and his resurrection, representing in his act, the health of soul and body, remedy all spiritual illness, Healing Addictions, repressing the passions, winning or weakening the temptations, communicating Tarab, confirming the nascent virtue, confirming the faith, strengthening the hope, igniting and dilating the charity.

In a passage described in the Bible, Jesus said: the sheep my dad given me, none will be lost.

Believe in these words and surely will have more peace of mind. In our short walk, we need to understand that all his wisdom lies in the subtleties of information travelling through time, added to the designs of the universe, which weaves transformation immortal intelligence, showing us his ample interests and forcing us through the multiple talents that we use with a certain boldness between the real and the imaginary that will gradually forming the superb panorama of our science, but which is briefer the reason that created it."

It's consistent, imagine that in the absence of a suitable intellectual training, faith degenerates in bigots. There are many ways to communicate with God and each must find his and after this, surely the purpose of life will become increasingly clear in your mind.

The religion has always been very important to our inner balance and there was a moment in our history, in which became necessary more empiricist, so we could better meet our real intrinsic values and we needed a great reference. One of the most delicate points in an attempt to understand the story of Jesus are his miracles, because at that time there was not a clear definition between the natural and the supernatural, but no matter, she is very fascinating, mainly by representing in our lives, the light at the end of the tunnel.

why does the science depends of the religion?

Because it would be impossible today to build airtight system science, in that all reality would be at your fingertips. She's far from being autonomous and defined by its methods, the nature of rationality, this is because the science itself is forced to lean on fundamental assumptions of religion. There are still many questions, which I'm sure we can take the ordered nature of the physical world, as well as the human capacity of this perception, without looking through the creator's rationality, theism.

In short, science can be anything but self-sufficient, because separating the religion of science so they don't fight each other, is only half the story. Postmodernist perspective none of them can claim superiority. The question of the future of the universe this connected directly the amount of matter, the universe has.

The combination of ordinary matter formed (protons, neutrons and electrons), matter and dark energy determines not only the dynamics of the universe (delayed or accelerated expansion), but also the geometry (cases in which it is opened, closed, or plan). This combination of ordinary matter, dark matter and dark energy define their geometry and, consequently, their final destination.

Maybe the future of our universe is in a fresh start with another (or, according to cyclic model, the same) Big Bang.

Now to know about the future of life within this universe, is a subject much more interesting...

IINFORMATIONS ABOUT NEW PUBLICATIONS

WWW.facebook.com.br/wmdias

Copyright © 2014 by

Wladimir Dias

www.ingramcontent.com/pod-product-compliance
Lightning Source LLC
Chambersburg PA
CBHW081130170526
45165CB00008B/2622